Deck Ahoy!

Primary Mathematics Activities and Games
Using Just a Deck of Cards

Janis A. Abbott

We hope you and your pupils enjoy using the ideas in this book. Listed below are a few of our other books which might be of interest to you. Information on these and all our other books can be found on our website: www.brilliantpublications.co.uk.

Published by Brilliant Publications
Unit 10
Sparrow Hall Farm
Edlesborough
Dunstable
Bedfordshire
LU6 2ES, UK

www.brilliantpublications.co.uk

The name Brilliant Publications and the logo are registered trademarks.

Written by Janis A. Abbott
Illustrations and cover illustration by Emily Skinner

© Text Janis A. Abbott 2015
© Design Brilliant Publications 2015
ISBN printed book: 978-1-78317-178-1
ISBN e-pdf: 978-1-78317-180-4

First printed and published in the UK in 2015

The right of Janis A. Abbott to be identified as the author of this work has been asserted by herself in accordance with the Copyright, Designs and Patents Act 1988.

Contents

Upper Key Stage 2 activities

Statistics

Teaching time

Introduction

Why should an ordinary deck of cards become an essential tool to teach maths?
Here are a few reasons:-
❖ It's inexpensive – cheap to buy; cheap to replace
❖ It's familiar – many homes have at least one deck
❖ It's dual purpose – leisure and learning
❖ It offers seemingly, endless possibilities
❖ It's easy and it's fun.

The suggestions which are included in this booklet are not intended to substitute the many excellent published mathematics schemes found in our schools. These games and practices are most effective when they are used to complement and extend the textbook exercises we use in the classroom.

Some of the exercises and games can be used prior to textbook work – for very young children or children with SEND.

Some can be used to introduce textbook work or as an alternative to the ubiquitous Mental Starter.

Some can be used to reinforce textbook work (further examples of the same task with individually generated numbers.)

They are great for homework (everyone at home can get involved) and there are no worksheets to photocopy or mark.

In next to no time, you and your pupils will find new and effective ways to practise, consolidate and extend their mathematical skills.

Many of the tasks lend themselves to written work, too. I usually ask children to write 10 examples in their books (labelling them A, B, C etc, because they are calculations of number). This not only consolidates their mental skills, but gives evidence of the learning and achievement that has taken place. Whilst written work is not always required, it should not be ignored either. A good balance of written and mental maths is easily achieved with *Deck Ahoy!*

Welcome to Deck Ahoy!

Activities are labelled by the key skill phrase or a game name and are divided broadly into three sections:

❖ Key Stage 1
❖ Lower Key Stage 2
❖ Upper Key Stage 2

On the contents list on pages 3–5, I have linked the activities only to the year in which a skill first appears in the National Curriculum for England, rather than provide references for every time the skill is repeated using larger numbers. For some of the activities, there are no direct National Curriculum references. I would include these activities in my lessons, regardless.

Please do not be restricted by these labels.

If *Deck Ahoy!* is introduced at Key Stage 1, then the activities will probably be met in the time scale suggested by the labels. If you are introducing *Deck Ahoy!* to a Y5 class then you may wish to go back to basics rather than start at Upper Key Stage 2.

There is no reason why able or gifted children should be confined to activities in their Key Stage. Move on!

Many of the activities are compatible with the 'Mathematics Passport' concept, adopted by numerous schools around the country, which arranges skills of number recall and calculation into several categories of learning (eg, Continents or Planets). Some also refer to Mathematical knowledge of Measurement (including time). All of these skills can be complemented by *Deck Ahoy!* It is a useful tool for pupils to assess each other and record their progress through their passports.

Additional activities can be found which are not included in the curriculum but which I have found effective and valuable and pupils have enjoyed.

From time to time, additional materials are useful or required. A3/A2 poster paper is essential for *Times tables mats* (page 15) and comes in handy when setting variations for each suit. Coloured counters or cubes and whiteboards and pens are needed for the later stages of *Clock solitaire* (page 53). Base 10 apparatus gets an airing, too. Other materials will creep into the scheme of things but nothing which is expensive or elusive, I am sure.

The more I teach, the more I realize that my own knowledge is neither complete nor infallible. So, please alter, amend, correct or dispose of anything you feel is inaccurate. And *please* add your own ideas as you find your feet, on deck!

Terminology

Most terms will be familiar but you say 'boot' and I say 'trunk', you say 'lunch' and I say 'dinner' etc. – so let me just explain a few and, forgive me now, for stating the obvious:

Complete deck	All 52 cards and the 2 Jokers
Closed deck	Cards face down, closed together, held in the hand or placed on the desk
Open deck	Cards face up, spread horizontally, held in the hand or placed on the desk
Faceless deck	A, 2, 3, 4, 5, 6, 7, 8, 9, 10
Digit deck	A, 2, 3, 4, 5, 6, 7, 8, 9
Zeros	The face cards (J, Q, K), placed face down
Twenties	The Jokers (can be used for any maximum number you require)
Suits	Hearts (red), diamonds (red), clubs (black), spades (black)
Shuffle	Mix'em up!
Stacked deck	Stack cards in numerical order, face up (in suits), lowest value at the bottom, highest at the top
Face down	No number or face showing
Face up	Showing number or face
Random and Rapid Fire! (**RRF!**)	This is the part the kids love the most. When they have practiced a skill to the point they are sure they can recall randomly and rapidly, shuffle the deck and turn over each card, one at a time, but rapidly as the child calls out the response. This is excellent for homework and parents love this, too. They can see for themselves, their child's achievement. Exercises that lend themselves to this will be punctuated with '**RRF!**'. (RAPID FIRE with a 'stacked' deck is the first step. Shuffle the deck when the pupil's knowledge is secure – **RRF!**. You will probably find that the kids will refer to both as RAPID FIRE – the word RANDOM seems to be an 'Adult Thing'.)
Statistics	Once called Data handling. There is a section devoted to Statistics on page 49.
Clock solitaire	This traditional game is the basis of activities devoted to teaching the measurement of *time*. It is fully explained in the final chapter, Teaching Time (see page 51)
Units	It appears the term 'units' is being replaced with the term 'ones' in the UK. When I am assured that every other nation in the world has followed suit, I too will embrace the new terminology. Until then…

Who has the deck?

In an ideal world, each child would have a deck of their own. In the past, I have known affluent parents, delighted to equip their child with a deck for school and another for home.

Recently, I have provided a class of 30 with a deck each, to use in school and take home, as their 'prize' for completing a reward chart. Schools may opt to provide each pupil from school funds or fund-raising events. Some may provide a 'class set' to be used in school only and be shared between classes. The solution will arise according to the needs of the pupils and the schools, I am sure.

If children provide their own decks, encourage them to buy cards with unusual backing pictures. If school provides the decks, mark each pack with different combinations of stars, stickers, etc. This makes life so much easier when children drop a card, leave their deck in the book corner, put it in someone else's tray by mistake... . These things can and will happen.

Be patient! The initial excitement (and ultimate waste of time which accompanies such excitement) will pass. Within a few weeks, your class will produce their cards at their desks with the minimum of fuss and set to work calmly and diligently.

Should a child forget their cards, or you do not have a deck for each child, remember – *half* a deck is better than none! Sometimes it is necessary to share it according to suits (so each child has one complete red suit and one complete black suit). Sometimes it is simply a matter of handing over roughly half the number of cards, with no regard to suit or value. (I always found that giving team points to the child who shared their deck, encouraged more children to bring in their own! Kids, eh?)

Finally, all these tasks will need to be introduced and modelled by an adult. However, it is surprising how quickly children not only recall the exercises, but quickly adapt to new numbers and calculations.

The exercises, tasks, games – call them what you will – can be individual, pair or group activities. The curriculum references are progressive but the activities within are not hierarchical.

Please trawl and hopefully find something to suit your needs. Make them your own. Make them work for you and yours.

I hope you find *Deck Ahoy!* at worst useful, and at best, inspirational and you and your pupils fan the flames of a burning desire to excel in mathematics.

Enjoy!

Key Stage 1

Suit sort

Use a complete deck, bar the Jokers. Select the Kings and place them in a row. Hold the remaining cards in a closed deck, face down. Turn over one card at a time and place it on the correct King. Say aloud the name of the suit/colour. Repeat until complete. **RRF!**

Extend to say the value and suit, eg 'the 6 of clubs', 'the 9 of diamonds' (Ace = 1, Jack = 11, Queen = 12, King = 13).

Odds/evens

Use a full, shuffled, closed deck and, turning over one card at a time, sort odds to the left, evens to the right, saying the value aloud and stating odd or even, eg '2 is even', '11 is odd'. **RRF!**

(Remember! Joker = 20)

Single digit recall/single digit +1

Using a closed digit deck, shuffle, then turn over the cards and say what they are.

For 'Single digit +1', turn over the cards and add one. Begin by saying the whole number sentence eg '4 add 1 equals 5'. Encourage children to use synonyms for add/equals.

Extension
The child simply calls out the answer (doing the calculation mentally). **RRF!**

Single digit -1

Using a closed digit deck, shuffle, turn over the cards and take away 1. Begin by saying the whole number sentence eg '4 minus 1 equals 3'. Encourage children to use synonyms for minus/equals.

Extension
The child simply calls out the answer (doing the calculation mentally). **RRF!**

Add on to 10

Using a closed digit deck, shuffle, turn over the cards and count on aloud until ten.

Order numbers to 10

Use a shuffled, faceless deck. Turn each card and place in a row, ordering 1-10. Ignore colour and suit. If the same number appears, put it to the bottom of the deck until the first row of 10 is complete. Then start a second row. Finish when all four rows are completed.

Extension

Order from 10 – 1.

Digits showdown

Use a digit deck (or a full deck if you want to include 10, 11, 12, 13 and 20 when children are able). With the deck closed, turn over two cards and place them side by side, face up on the desk. Describe the greater value in comparison to the lesser value (eg '7 is greater than 4'). Place the two cards to the bottom of the deck and continue. (This can be the basis of a written task using >, > and =.)

It is easy to see how you can use reverse language and synonyms. It is also easy turn this into a competitive game between two or three children, like 'Snap' but the highest value wins and collects all the cards. If two cards of the same value appear, players turn over another card each, increasing the 'pot'!

Showdown, like *Old maid* (see page 12), is open to infinite variations. You will encounter it many times in this book, in many forms.

Single digit +10

Using a closed faceless deck, shuffle, turn over the cards and add 10. Begin by saying the whole number sentence eg '4 add 10 equals 14'. Encourage children to use synonyms for add/equals.

Extension

The child simply calls out the answer (doing the calculation mentally). **RRF!**

Add on to 20

Place a Joker on the desk, face up to represent 20. Use a closed, full deck. Turn over the first card and count on to 20.

— —

20 Countdown

Place a Joker on the desk, face up to represent 20. Use a closed, full deck. Turn over the first card and count backwards from 20 until you reach the value of the card.

— —

Old maid

Use a faceless deck **with the addition of** the Queen of Spades. Deal the cards face down amongst a minimum of 3 and a maximum of 5 players. Players take turns to put down (face up) a pair of any number – saying aloud '2 threes' or '2 tens', etc.

When all the pairs have been placed and stated, the players take turns to take a card from the player to their right. If a pair can be made with the newly taken card, the pair is laid on the table. Play continues until most players have no more cards and one player is left with the **Old maid** (the Queen of Spades).

There are many variations of this game, so please make your pupils well acquainted with it!

Deck Ahoy!

Ordinal numbers

Use only one suit at a time, initially, then progress to two (one red, one black) and finally, all four.

Stack the suit, hold as a closed deck and turn over each card, saying the ordinal numbers instead of the digit value, ie 'first', 'second', 'third', etc.

Once the language is established, shuffle the cards and repeat, but leave each card face up on the desk, for the child to see and position in the correct order. After the K 'thirteenth' there would be a gap until the Joker 'twentieth'. To fill this gap, use a red A (as 10) and the other digits (black) to represent the units, ie A4 for 'fourteenth' etc.

— —

Ordinal positions

Use a full deck and turn the cards face up, in two or more horizontal rows. Then ask simple questions (eg 'which card is in 5th place?'). You may accept a simple pointing of the finger or expect a verbal description (eg 'the six of hearts') according to the child's ability. Extend to swapping eg 'Move the seven of clubs to eighteenth place'.

— —

Number bonds to 10

Use a faceless, shuffled deck and four *zero*s. Select A–10, any combination of colour and suit, and place in two horizontal rows of five cards, (A, 2, 3, 4, 5) (6, 7, 8, 9,10). Shuffle the remaining cards. Hold the closed deck face down and turn over a card. Place it on top of the digit it bonds with to make 10, saying aloud e.g. '3 and 7 make 10' *or* 'zero and 10 equals 10'. (When 10 appears be sure children place it at a distance as they say '10 and zero' and do not place the 10 on top of another 10 – which, of course, would make 20.

If a turned card cannot be placed on its bond, return it to the bottom of the deck. It will reappear and eventually be placed with its bond. Continue placing the cards on top of the digits they bond with until the whole deck is paired.

Old maid number bonds to 10

Play as *Old maid* (see page 12) but place cards down in number bonds to 10, saying the number sentences aloud e.g. '6 and 4 makes 10'. Extend both activities to multiples of 10; bonds to 100; multiples of 100; bonds to 1000.

Old maid doubles to 20

Play as *Old maid* (see page 12) but calculate the pair as a double saying aloud eg 'double 7 is 14' or '2 times 9 equals 18' etc.

Old maid doubles and number bonds to 20

Play as *Old maid doubles to 20* (see above), but add to make 20 saying aloud, eg 'double 7 is 14, add 6 is 20' or '2 times 3 equals 6, add 14 equals 20' etc.

Variations and extensions
Add J, Q and K (11, 12, 13), saying aloud eg '2 times 13 (K) equals 26, take away 6, equals 20

Use multiples of 10, 100 etc.

double 7 is 14, add 6 is 20

Deck Ahoy!

Creating a times tables mat

This activity requires two people in the first instance, usually an adult or a more able child and the 'learner'.

I usually get the whole class to make their times tables mats at the same time. The differentiation can be accommodated after the initial drawing. You will need A3 paper and pencils, coloured pencils/felt-tipped pens.

Demonstrate **before** you allow children to produce theirs.

When children count on their fingers, they **usually** do so in a common pattern. A right-handed child tends to start with their right thumb (1, 2, 3, 4, 5) then continues with the left thumb (6, 7, 8, 9, 10).

A left-handed child, tends to start with their left thumb (1, 2, 3, 4, 5) and continues with their right thumb (6, 7, 8, 9, 10). Establish these with your class before you begin to draw the mats.

Place your hands on the mat side by side (thumbs in the middle, fingers slightly spread apart) and draw around them, one at a time.

It is good practice to encourage children to do this for themselves no matter how much they protest or how untidy their first efforts may be. That will demonstrate to them, the ease with which they can replace a forgotten, lost or damaged, times tables mat.

They can have one at home, one at school, one at Granny's etc. If you draw for them or even worse, produce a template, you will simply suggest that this is an activity for school and it does not extend beyond the classroom. An underlying principle of **Deck Ahoy!** should be 'you can practice maths anywhere!'

Once the hands are drawn, demonstrate numbering the mat for a right-handed child. Using a pencil, write the numbers where the finger nail would be, starting with the right thumb moving right (1, 2, 3, 4, 5) then going to the left thumb and moving left (6, 7, 8, 9, 10).

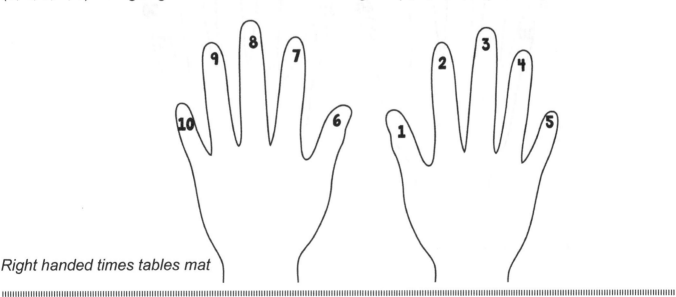

Right handed times tables mat

Then demonstrate for left-handed children, starting with the left thumb and moving left (1, 2, 3, 4, 5) followed by the right thumb, moving right (6, 7, 8, 9, 10).

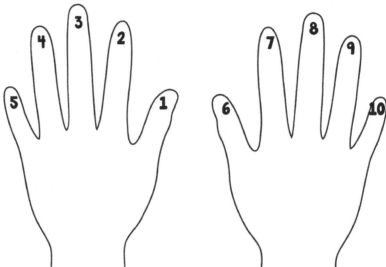

Left handed times tables mat

If any child has not yet grasped the concept of **odds and evens**, the numbers could be written in alternating colours to reinforce their knowledge.

Now … for the mathematics. Write the products of the times table you wish the children to learn, above the corresponding digit. If the class are learning the 10x table the products would be 10, 20, 30 etc above fingers 1, 2, 3 etc. (So, naturally, if the children are learning the 6x table, the products 6, 12, 18 would be written above fingers 1, 2, 3.)

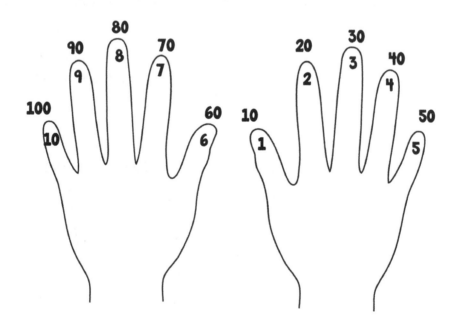

Sample times tables mat for 10x (for right hander)

Deck Ahoy!

Times tables recall

Using a closed, stacked deck, turn the cards slowly giving the child ample time to repeat after you eg (3x table):

❖   3 times 1 equals 3
❖   3 times 2 equals 6, etc.

Ensure the child taps the corresponding finger up and down on their *Times tables mat* whilst reciting the maths. There seems to be a very real connection between the physical action and the recall of the times tables fact.

If you continue with the stacked deck, the child will go through the times table *four* times by the end of the deck. Now try going through the deck, without the mat. Then shuffle the deck and try **RRF!**

If progress is less than rapid, practice the simpler, routine exercises more frequently. Attempt **RRF!** only when success is more likely than failure!

(Once the child is able to recall rapidly and randomly you may wish to introduce the J, Q and K for 11x, 12x and 13x, as the National Curriculum requires recall to 12x. The 20x can be included to enable more able pupils to work with the concept of 'double and multiply by 10'.)

When the child is ready to move onto the next times table, just write the new products in a new colour above the previous numbers!

It's that simple! If the sheet looks too crowded, too untidy – draw a new one on the back or on a new piece of paper. So easy, so effective!

Before long, the children will come straight into class and practice **RRF!** as Solitaire – working on their individual times table target. It is a truly encouraging sight – an enthusiastic pupil, flicking through a deck of cards as fast as they can (or alternatively, in pairs, so they can test each other).

All digits number bonds

This may be the most important part of addition and subtraction and children don't always understand how important it is in mental maths! Persevere with this and the reward will be speed and accuracy.

Using a stacked, digit deck, turn over each card and chant the pairs of numbers that bond to make the value of the card, in ascending order eg:

(5) '0 and 5', '1 and 4', '2 and 3', '3 and 2', '4 and 1', '5 and 0'

(Note: KS1 and Y3 pupils need to chant the repetitive pairs. They don't always make the connection without it, so don't stop at '2 and 3'.)

— —

'007' Ludo ('bonds, games bonds')

Ideally, this is a game for only two players so that concentration does not wane. Any more and children tend to drift off task when they are several 'turns' away from the action.

Use an ordinary Ludo board from any games compendium or a free, printable web image. Use *only* two counters/'men' of the same colour per player. Set up the counters in adjacent bases (not diagonally opposite) or the players may never be in a position to 'attack' each other. Use a shuffled, faceless deck.

The first player, turns over one card and must split its value into a bond **and declare it**, **before** they can move their counters. 0 is not an option. Both counters must move. If a player turns over an Ace (ie '1 + 0') a choice of rule can be made – either the player will miss a turn **or** the card can be ignored and another turned over. Children tend to opt for the **miss a go** rule – much more fun!

If their declaration is mathematically incorrect and the **other player** challenges it, the first player loses their turn. This is another **miss a go** rule that children love.

If their declared number bond is correct, they move both of their counters accordingly.

As play continues, a player may land both their counters on the same space. That is 'safe'.

A player 'attacks' their opponent by selecting a number bond that allows one of their counters to land on top of their opponent's, thereby sending it back to the 'start'.

The first player to get both of their counters 'home', wins.

Variation
'007' Ludo deluxe: play as before but use all the cards, with J, Q, K and Joker corresponding to 11, 12, 13 and 20.

Take that snakes 'n' ladders

Use a standard board from any games compendium or a free, printable web image. Use a shuffled, faceless deck.

Each player has only one counter. (To ensure full concentration it may be best to impose a maximum of three players in the game.)

The first player turns over the first card and declares a number bond. If it is correct, the player moves their counter one part of the bond but gives one other player the other part of the bond eg (8) '5 and 3'. The first player may move himself 5 and the other player must move 3. (The players must move the declared spaces. No-one may refuse.)

As the game progresses, pupils quickly learn the strategy of splitting their number into bonds that will allow them to progress up the ladders and send their opponents down the snakes ('Take that'.)

Miss a go rules apply as in *'007' Ludo*.

Variation

Take that snakes 'n' ladders deluxe: play the game as before but use a full deck with J, Q, K and Joker corresponding to 11, 12, 13 and 20.

'007' Ludo and *Take That Snakes 'n' Ladders* are wonderfully effective at securing number bonds, essential for quick mental maths calculations. They are both excellent for homework! Parents, grandparents and childminders love them too. Many children have boards and counters at home.

Tip

Laminate several copies of both boards so children can use them for homework or lessons. They often choose to play even in 'Golden Time'!

Add the deck

Very simple in principle, this activity gives endless variety and differentiation. Use a faceless deck, shuffled and closed. Encourage pupils to use their knowledge of number bonds and number bonds to 10, to add quickly and accurately. It is more accurate and quicker for mental maths than counting on their fingers!

For example: 6 + 9 = 6 + 4 (10) + 5 = 15

7 + 5 = 7 + 3 (10) + 2 = 12

Variations

❖ Turn over 2 cards and add them (maximum 20).

❖ Turn over 2 cards and add them. Turn over the next card and add to the previous total. (Maximum to be set according to ability eg 30, 50, 100.)

❖ Turn over 2 cards and use as a 2-digit total – continue as before until the total is reached

❖ Add the complete deck in single digits.

❖ Competitive – each player turns over two cards, calculates the total and the player with the highest total collects all four cards.

❖ Teacher led/peer assessed – the teacher decides how many cards each pupil will turn over. Each child totals their cards, writes their answer and then swaps cards with a partner for assessment.

Extension

Extend the challenge for more able pupils by including the face cards and Jokers (11, 12, 13, and 20)

Simple subtraction

Turn over two cards. Place highest value to the left and read the two cards as a number sentence eg if the cards were 9 and 7: '9 take away 7 is 2' (9 − 7 = 2). Extend to 2-digits take away a single digit.

Simple inverse operations

Use a full deck. Turn over two cards but use each as an individual number. Verbally model different operations using the correct answers and language eg 6, 9: '6 add 9 equals 15'.

From that one statement, generate three others using inverse and commutative laws ie:

- ❖ '6 add 9 equals 15'
- ❖ '9 plus 6 equals 15'
- ❖ '15 minus 9 equals 6'
- ❖ '15 take away 6 is 9'

Use synonyms for the operations so children become familiar with them all but ensure that every time a number sentence is generated, it is followed by 3 others!

Variations and extensions

Repeat for multiplication and division.

Extend to one 2-digit number and one single digit, then 3-digit, etc.

As the pupils progress through Lower Key Stage 2, they should be able to calculate a number sentence using several digits, by calculator, but still express the inverse operation and commutative law statements, eg $3,426 \times 516 = 1,767,816$

The ability to calculate this is not as important as the ability to find the missing number:

$1,767,816 \div \boxed{} = 516.$

As the pupils progress through Upper Key Stage 2, they should be able to complete similar tasks using decimals and multiples of 10 or 100.

It's a simple skill but often forgotten after the initial work on number bonds to 10 or 20!

Lower Key Stage 2

Multiples showdown

For children who have learned the basic tables (10x, 2x, 5x), use Red A, 2, 5,10 (shuffled) and Black A–10 (shuffled). Play in pairs. Each child turns over 1 red card and 1 black card. Each child calculates the product of their own cards. The pupil with the highest product collects all four cards eg 4 x 10, 5 x 7: the pupil with 4 x 10 takes all four cards because 40 is greater than 35! Play a set number of rounds or a set time and count who has the most cards.

Because each calculation includes a red card (A, 2, 5, 10) and a black card (A–10), pupils will experience the commutative law of multiplication, introducing other times tables facts, eg: 5 x 6 = 6 x 5; 2 x 7 = 7 x 2.

— —

Division facts recall

This is not as natural an activity as times tables recall (and the language is *so* important) but once learned, an easy and effective method. Use the times tables mat as a reference chart. Use a shuffled, faceless, closed deck.

Turn the cards one at a time saying aloud eg, if the card turned over is 7, 'something ÷ 10 is 7'. If the child refers to the 10x product above their 7th finger on their times table mat, the answer should be 70! (Say aloud the number sentence '70 divided by 10 is 7.')

It may appear cumbersome at first, but the children pick it up quickly!

— —

Factors showdown

Use a full deck. Remove the 10s and Jokers. Shuffle the deck. Face cards are played face down as *zero*s. Each child turns over two cards and arranges their place value to make their number divisible by 2, 5, and 10. The player, whose number has the most factors, collects all four cards. If there is a draw (tie), the cards are left in the 'pot' and added to the next round. For example:

Round 1	Player 1	Player 2	This is a **draw** because 75 has 5 as a factor and 38 has 2 as a factor
	57 (75)	38 (83)	
Round 2	35 (53)	K3 (30)	30 wins because it has 2, 5 and 10 as factors; 35 has only 5 as a factor

Play a set number of rounds. The pupil with the most cards is the winner. This activity will help children to practice and consolidate their knowledge of place value, as well as times tables.

Deck Ahoy!

Rapid fire 100s and 1000s

Use a shuffled, digit deck to generate 3-digit numbers for pupils to read aloud. Children progress very quickly reading and partitioning bigger and bigger numbers. Year 3s can read 7 digits in no time at all and love the challenge of questions such as (7 546 921) 'what is the value of the 4?' 'what is the largest number you can make with these digits?'

Written tasks are simple, effective and 'unique'. The number of digits should be set by the t according to the pupil's ability. The number can be written in numerals or words.

— —

Place value - ordering/comparing

Use a shuffled digit deck and turn over three cards. Write the 3-digit number as it appears, in the middle of the line. Write the smallest number achievable to the left; write biggest number achievable to the right.

Tip
Because the written work is based on numbers only, encourage children to use letters rather than numbers to label each example, when writing in their books, eg:

(382)   A   238   382   832

— —

Multiplying by 10

Single digit numbers times 10. Use a shuffled, digit deck. **RRF!**

Then introduce J, Q, K (11, 12, 13) then the Jokers (20). Return to a digit deck before moving on to 2-digit and 3-digit numbers, for more able pupils. **RRF!**

The next 10/the 10 before

This activity is a good introduction to rounding. Use a shuffled, faceless deck. Turn over two cards and read as a 2-digit number, then say the next 10, eg if their number is 26, they say 30. Concentrate on the *next 10* until secure.

❖ If 10 is turned over as the first card, do not turn over another. (It is important to have 10 in the deck so the pupils recognize that 0 is the ten before 10!)

❖ If 10 is turned over as the second card, return it to the bottom of the pack for later and use a single digit. (It is important to encounter some single digits so that pupils recognise the next 10 after 1 is 10!)

Then concentrate on the *10 before*. The *10 before* is so important for children when learning to round down. So many make the mistake of thinking 34 is rounded down to 20 or 81 is rounded down to 70. It is worth spending time on this – it makes a great homework task!

Finally combine the two, if desired, and use the colour of the last card to determine whether to call the *next 10* or the *10 before*. **RRF!**

- -

Rounding to the nearest 10

Having taught and consolidated rounding up and rounding down, use a shuffled, digit deck to generate 2-digit numbers. Round up or down to the nearest 10. Extend more able pupils by rounding 3-digit numbers to the nearest 10 or the nearest 100. **RRF!**

- -

Complements to 100

Use a shuffled, digit deck and turn over two cards to make a 2-digit number. Calculate the number bond to make 100, for example:

47 + 3 = 50   + 50 = 100   47 + 53 = 100

(Children may need reminding to start with the units and bond to the *next 10* first.) **RRF!**

Extend more able pupils to **Complements to 1000** by using 3-digit numbers.

Deck Ahoy!

Complements showdown

This is a game for two or three players. Use a shuffled, digit deck to generate 2-digit or 3-digit numbers, according to the pupils' abilities. Set a total for the complement eg 100, 1000, 800, 1500.

The first player to call out the correct complement 'wins' the cards. Complete the deck. The player with the most cards is the winner.

Alternatively, showdown can be scored with counters and the player with the most counters when the deck is completed, wins.

— —

Add the deck/scrub the deck

At Lower Key Stage 2 pupils are introduced to more formal written methods of addition and subtraction. The range of abilities in a Y3 or Y4 class can be very wide. Some are able to see their written calculations in their mind's eye and make mental calculations with ease. For the more able pupils who seem to have grasped the addition of two 2-digit numbers or 2-digit and 3-digit numbers (in any combination) with few difficulties, the following activities can be used:

❖ *Add the deck*: use a digit deck to generate a 2-digit number. Repeat. Add them together. Generate another 2-digit number and add to the previous total. Continue until the deck is completed!

❖ *Scrub the deck*: use a digit deck to generate a 2-digit number. Subtract from 101. Generate another 2-digit number and subtract from the previous answer. Continue until 0 is reached. Repeat.

However, for some pupils it seems to be a mighty leap and I have found that a combination of cards (to generate numbers) and base 10 apparatus helps many pupils to understand and consolidate their skills of carrying and exchanging (see below).

— —

Add the deck with base 10 apparatus

Base 10 apparatus is used in KS1 to give children concrete examples of their numbers. It is worth a mention in KS2 because of the new skills of carrying and exchange which sometimes need the reinforcement that base 10 apparatus can offer.

❖ Use a digit deck to generate a 2-digit number. Use base 10 apparatus to construct the number. Repeat. Add them together, tens in one set, units in another. If the units number more than 9, convert to 10 (ie carry 10!) If the tens number is more than 9, convert to 100 (ie carry!)

❖ Start again with a new pair of 2-digit numbers.

❖ Repeat!

This page may be photocopied for use by the purchasing institution only.

Scrub the deck with base 10 apparatus

❖ Use a digit deck to generate a 2-digit number. Use base 10 apparatus to construct the number. Repeat. Subtract. Ensure that the larger number has enough units by exchanging a ten for ten units, when necessary.

❖ Start again with a new pair of 2-digit numbers.

❖ Repeat!

— —

Darts

This is another popular game, for more able pupils who are able to double and treble numbers up to 20. It certainly helps to have a printout of a dartboard for pupils, for reference but the game is completely conducted by the turn of the cards.

Initially, pupils may add their scores to be the first to reach a total of 101, 301, or 501. (Set a winning score according to ability!) As pupils improve their skills with **Scrub the deck** (see page 25), they could begin with 101, 301 or 501 and subtract, as they would in a real game of darts.

Use a full deck. Turn over a card. Red = single value. Black = double value. *Unless* it is Clubs – treble value! Black Jokers are always treble 20! Subtract from starting score. Repeat.

There are many possible variations, eg turning two cards and adding (to give scores of more than 13 before doubling or trebling) or repeating twice more as three 'darts' to give a much higher total before subtracting.

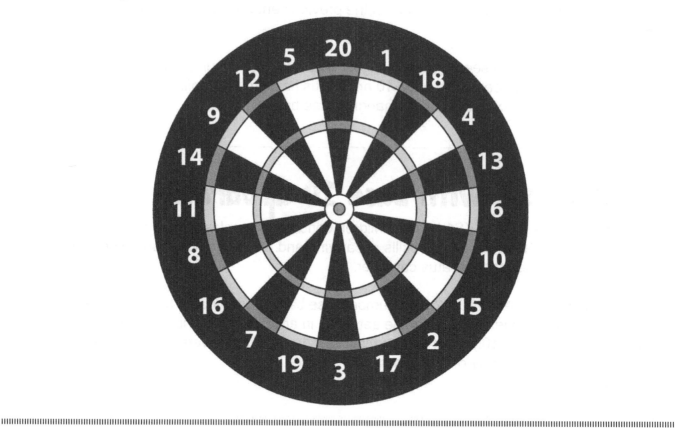

Deck Ahoy!
© Janis A. Abbott and Brilliant Publications

Divisibility tests

When pupils are able to recognize multiples of 2, 10 and 5, introduce divisibility tests. Use a shuffled digit deck. Turn over two cards. Read as a 2-digit number. Concentrate on the number in the units column to decide if the whole number is divisible by 2 (even), 5 (5 or 0) or 10 (0).

— —

Divisibility by 3

Use a shuffled, digit deck. Turn over two cards and place them side by side. *Read* aloud as a 2-digit number. Now add the digits together. If the sum is a product in the three times table, the 2-digit number is divisible by 3.

— —

Divisibility by 4

Use partitioning and halving to find half of the number and if the answer is *even*, it is divisible by 4.

— —

Divisibility by 8

Use partitioning and halving to find half of the number and if the answer is *even*, it is divisible by 4. Partition and find half, again. If the answer is *even*, again, the original number is divisible by 8.

— —

Divisibility by 6

If the number is *even and* divisible by 3, it is divisible by 6. Use a shuffled, faceless deck to generate 2-digit or 3-digit numbers to test for divisibility by 4, 8 or 6. (Divisibility by 9 and 7 will be introduced later and divisibility by 8 will be repeated there, too!)

— —

Multiples and factors

Play the *Showdown* games as before (see page 22) but extend to the 3x, 4x, 6x, and 8x tables.

Proper fractions

It is very easy to generate simple fractions. Use a faceless deck, initially. Later, extend to using a full deck with the usual values for the face cards and the Joker. Turn over a card and use its value as a denominator with 1 as the value for the numerator.

This first stage is essential for consolidating the language of fractions and pupils love to do this as **RRF!**

- -

Adding unit fractions to make 1

Generate a fraction as above, for example: $\dfrac{1}{6}$

Then add the complement to equal 1:

$\dfrac{1}{6}$ $\dfrac{5}{6}$

This activity may be coupled with drawing quadrilaterals on squared paper to provide a graphic example with shaded and unshaded areas.

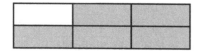

- -

Subtracting unit fractions from 1

Try an inverse operation! Start with 1 and subtract the fraction generated by turning a card from a faceless or full deck:

$$1 - \frac{1}{6} = \frac{5}{6}$$

The same graphic representation can be used to establish the language of addition or subtraction with fractions ie 'one sixth and five sixths equal one' or '1 take away one sixth equals five sixths' etc.

Deck Ahoy!

Counting in fractional intervals

Generate a denominator by turning over a card from a faceless deck. You may wish to limit the cards to the numbers 2, 3, 4, 5 and 10 or even 2, 3, 4, depending on the ability of the pupils.

Turn over another card as the starting number. Count from the starting number to the next whole number, in the fractional interval, eg if the first card turned over is 3 and the second is 2, the factional interval is: $\frac{1}{3}$ and the child counts 'two, two and one third, two and two thirds, three'.

This is the most valuable exercise in the teaching of fractions! It ensures that children not only know the language of fractions but that when counting, they will never say **three thirds** or **four quarters**, they will say the next whole number!

Extend by generating a starting number **and** a finishing number. Initially always start with the lowest value. First count forwards 'two, two and a third, two and two thirds, three'. Then introduce counting backwards, eg: 'three, two and two thirds, two and a third, two'. (We take it for granted that children will say 'a third' when they write $\frac{1}{3}$ or that they will understand that when they hear 'a third' it means 'one third'; this activity irons out all the little creases in their understanding of fractions. Simple but effective!)

Next … if the starting number is greater than the finishing number, pupils will know to count backwards.

It is possible to have 15 pairs of children working independently with a deck of cards each, counting forwards or backwards in different fractional intervals. Noisy, but very rewarding!

— —

Converting tenths to decimals

Using 10 as the denominator, generate a numerator by turning over a card from a digit deck. Model the conversion of tenths to decimal tenths, eg $\frac{3}{10}$ is 0.3 etc.

All the previous activities of adding/subtracting and counting in intervals can be applied to tenths and decimals.

RRF! of decimals and tenths is very popular!

Red = tenths
Black = decimals

So 6 of diamonds = 'six tenths' but 6 of spades = 'zero point six'.

Roman numerals

RRF! with a faceless deck or a full deck is a great way to consolidate numbers less than 20.
More able pupils will enjoy challenging themselves or each other to create 2-digit or even 3-digit
numbers or more:

| Turn over 5 | Answer is 'V' | Turn over 9 | Answer is 'IX' |
| Turn over Q | Answer is 'XII' | Turn over 28 | Answer is 'XXVIII' |

The kids love it, especially when they get to 3-digit numbers or they are asked to give their
birthday in Roman numerals.

Any of the previous activities can be adapted to Roman numerals, even fractions. They can
convert to Roman fractions, by writing the turned card's value as a denominator in Roman
numerals, eg 10 becomes $\frac{1}{x}$.

— —

Length, perimeter and area

Most maths schemes deal with these concepts very well but should you wish pupils to generate
their own numbers, here is a simple method or two.

Use a shuffled, digit deck and place three cards, face down in a horizontal line. Demonstrate that
the line is not to scale. Tell the pupils which unit of measurement you wish them to convert, eg cm
to mm, m to cm or km to m, so 6 becomes:

| 6cm | = | 60mm |
| 6m | = | 600cm |

Turn the card face down and repeat, turning over one of the other cards. Use vertical lines for
variety.

Extend by turning over 2 cards eg 37 = 3m 7cm etc. Increase the challenge and extend the
knowledge by varying the task according to the colour, eg the first unit of measurement is metres.
If the card is red, convert it to centimetres. If the card is black, write in decimal form, eg 3.7m.

— —

Perimeter

Form a rectangle from two horizontal rows and two vertical columns of 3 cards. (Explain that the
model is not to scale!) Turn over one card in one horizontal row to represent the length. Turn over
one card in one vertical column to represent the height.

Calculate the perimeter (2l + 2h) and record the answer in the desired unit of measurement. Turn
the cards face down and repeat using different cards. Then return all cards to the bottom of the
deck and repeat with new cards.

Deck Ahoy!

Area

Do as for **Perimeter**, but multiply one card value by the other and record the answer as a unit of measurement. This is an excellent introduction to squared numbers; turning only one card on any side and using its value as both the height and the length (Remind children that the model is not to scale!)

Eventually, the quadrilateral shape will not be required and tasks on Perimeter or Area can be set simply by saying, eg 'turn over 2 cards of single digits and calculate a perimeter':

(9 & 5)   18 + 10 = 28m

Or 'turn over one card and use its value as the length of a square'. Draw the square.

— —

Square numbers 1-100

Having learned most of their multiplication and division facts to 12x, most children will be able by now to recognize pairs of factors. They will also begin to recognise that most sets of factors are even in number. A set that is odd, therefore must have a square number. For example:

Number	Factors	Number of factors	Odd or even number?	Square number?
12	(1, 2, 3, 4, 6, 12)	6	even	no
4	(1, 2, 4)	3	odd	yes
16	(1, 2, 4, 8, 16)	5	odd	yes

Using sets of factors is also a good way to introduce or reinforce **prime numbers**. A set of factors with only 2 numbers represents a prime number: 1 and the number itself eg 13 (1, 13); 59 (1, 59).

Start with a stacked deck and turn over each card slowly, allowing the pupil to say the number sentence aloud, eg '1 times 1 is 1' or '3 squared is 9'. Encourage synonymous number sentences.

Shuffle the deck when the pupil's knowledge is secure. **RRF!**

Extension

This can be extended to cubed numbers for more able/Upper Key Stage 2 pupils: $2 \times 2 \times 2 = 8$; $3 \times 3 \times 3 = 27$.

Partitioning and doubling

Knowledge of place value and the 2x table should enable many children to master this very quickly with a deck of cards and some nifty hand signals.

Use a shuffled, digit deck. Turn over two cards and read them aloud as a 2-digit number, whilst bringing their hands together.

Partition (and move the hands apart) into the value of the tens (indicated by their left hand) and the value of the units (indicated by their right hand).

Double the tens first (turning down the left palm) then the units (turning down the right palm).

Recombine (bring the palms together) by adding the two new values. For example:

'Double 31. **Partition**: 30 and 1 – double 30 is 60; double 1 is 2. **Recombine**: 60 and 2 is 62. Double 31 is 62.'

If doubling crosses the tens barrier, simply add it as 10, for example:

'Double 47. **Partition**: 40 and 7 – double 40 is 80; double 7 is 14. **Recombine**: 80 and 10 and 4 is 94. Double 47 is 94.'

Children pick this up very quickly and enjoy using their hands to simulate the calculation but it is *imperative* they use the correct language in the correct order.

Some will make the cardinal error of doubling the first digit instead of its value in tens! Establish good practice and they will be doubling multiple-digit numbers in no time!

This extends to written form very easily, allowing each pupil to work at their own pace, independently (because they generate their own numbers) without copying! For example:

	29				126	
	20	9		100	20	6
D	40	18	D	200	40	12
	40 + 10 + 8			200 + 40 + 10 + 2	
	58			252	

The addition of the face cards will introduce 0 into the generated numbers but the children must be instructed to turn a face card down so only the backing pattern is visible. If the face is visible, that would give it a value of 11, 12 or 13 which is not appropriate when partitioning.

Deck Ahoy!
© Janis A. Abbott and Brilliant Publications

Partitioning and halving

Introduce this as a whole class activity, ensuring that children can recognise the difference between an odd and an even number (no matter how great the value) by concentrating on the last digit.

This is a good time to introduce or revise the different ways in which an odd number can be turned into an even number, mathematically (eg 'double it', 'add 1', 'subtract 1'). Establish that 'divide by 1' or 'divide by 2' will not.) Select 'subtract 1' as the basis of 'remainder 1' when finding half of an odd number.

Use the same language and the same hand actions as before when doubling but substitute the words 'half of' for the word 'double'. For example:

	48				47	46	r1
	40	8				40	6
H	20	4		H		20	3
	20 + 4					20 + 3	
	24					23r1	

So,   'half of 48 is 24'   but   'half of 47 is 23r1'

Note: some children may need a lesson or a 'mental starter' devoted to finding half of all multiples of 10 up to 100 to ensure that they know half of 10, 30, 50, 70 and 90 will be 5, 15, 25, 35, and 45 respectively. Many discover the pattern when doubling; some do not.

- -

Multiples and factors

Play the **Showdown** games using the 7x and 9x tables (see page 22).

- -

Proper fractions

Use a shuffled, digit deck. Turn over two cards and align them vertically, always placing the lowest value at the top, to generate a proper fraction. Return the cards to the bottom of the deck and repeat.

Equivalent fractions

Use a die to generate a third number. Multiply the Numerator and then the Denominator by the number on the die, to generate an equivalent fraction.

Record both before returning the cards to the bottom of the deck. For example:

Number on die is 3 $\qquad \frac{2}{5} = \frac{6}{15}$

— —

Converting fractions to decimals

Introduce the conversion of fractions into decimals with the most familiar and their decimal equivalents in currency or length. For example:

$\frac{1}{2}$ = £0.50 $\qquad \frac{1}{4}$ = 0.25m $\qquad \frac{3}{4}$ = 0.75

Demonstrate the use of the calculator (set to 2 decimal places for Lower Key Stage 2) to produce the decimal equivalent (ie 1 ÷ 2= 0.5).

Then, using a shuffled, faceless deck, turn over two cards and arrange them as a proper fraction eg $\frac{5}{7}$. Demonstrate converting to a decimal with the use of the calculator ie 5 ÷ 7= 0.71.

Set pupils the task of discovering the decimal equivalents of the proper fractions they generate with the deck or equivalent fractions they calculate by introducing a multiple, for example:

$\frac{5}{7} = \frac{20}{28}$ $\qquad$ 20 ÷ 28 = 0.71

Measures and decimals
Compare the notation of different measurements and their number of decimal places, for example:
cm   1cm 2mm = 1.2cm
m   1m 52cm = 1.52m
km   1043m = 1.043km

At any point in KS2 these concepts can be introduced and consolidated by generating numbers for converting measurements and also for **RRF!** eg turning two cards from a digit deck to make a 2-digit number. If the last card is Red, read as cm. If the last card is black, read the number as m. 56 (6 red) '56 cm'; (6 black) '0.56m'.

This is easily extended and adapted for bigger numbers and varying measurements of length, mass and volume.

Deck Ahoy!
© Janis A. Abbott and Brilliant Publications

Money

Most pupils are familiar with coins and notes up to £10 by the end of KS1. Calculating can be fun with *Deck Ahoy!* simply because it gives scope for many more examples and a mixture of operations. It is best introduced once pupils understand that any amount of money can be written to 2 decimal places. Once they have begun to write numbers of pence or pounds and pence to two decimal places, *Deck Ahoy!* can offer all sorts of possibilities.

Using a digit deck, turn over 2 cards. Say the number as pence. Write the number as pounds and pence. 35 pence is written as £0.35.

Repeat until secure. Begin to generate two or more numbers of pence. Add them. This allows some answers greater than 100 pence, eg 78 pence and 64 pence written as:

```
  £0.78
+ £0.64
  £1.42
```

Extend by adding three or more addends.

Subtract by setting a starting number eg one pound £1.00. Generate a number of pence as before and write as pounds and pence. One pound take away 62 pence is written as:

```
  £1.00
- £0.62
  £0.38
```

The method of calculating the answer is open and should be according to ability. The example is only intended to show the notation to two decimal places.

It is now easy to set all sorts of activities and tasks using the decimal notation of money: adding prices, selecting notes and coins to match totals, finding change, finding multiples (multi-buys), comparing single and multi-pack purchases, etc.

Missing number problem solving

This topic often comes up in the optional SATs papers. This activity will give your pupils enormous confidence and accuracy with solving problems involving missing numbers.

Use a digit deck and model the addition of two 3-digit numbers eg 243 + 466.

Demonstrate converting the horizontal number sentence into a vertical calculation.

```
    243
+   466
    709
```

Construct the whole sum in cards, using a face card (face down) for the 0 in 709.

Turn one digit in each number, face down. Make sure that each turned card is in a different column from any other.

Demonstrate how to use our knowledge of number bonds and carrying to rewrite the correct calculation in spite of the 'missing numbers'.

Task pupils with repeating the activity in pairs. Generate a 3-digit number, then another below it. On a whiteboard, calculate the answer and construct the answer with their cards, below the sum line (a pencil or a ruler placed horizontally will do.) Turn one digit in each number face down. Make sure that each turned card is in a different column from any other. Clear the whiteboard. Leave the missing number problem for another pair to solve. Go and solve the missing number problem from another pair! Always replace the cards in their original position after solving the problem. Allow pupils to roam and solve as many problems as they can. They can record their calculations or just enjoy the activity.

For Lower Key Stage 2 this is usually an extension task for more able pupils. For Upper Key Stage 2 this should be a task for all at some stage. Find examples from the Optional SATs papers – especially those written in horizontal form. Encourage pupils to convert to vertical form if they have the skills to carry.

This is easily adapted for subtraction, multiplication and division and there are many online websites which have free printable sheets with examples to recreate using cards. Noisy, but effective!

Deck Ahoy!

Positive and negative numbers

Use a shuffled, faceless deck. Establish the black cards as positive numbers; the red cards as negative numbers. Turn over two cards and demonstrate the number sentence and rules of negative numbers, for example:

B5 + B6 = 11	R5 + R6 = -11	B5 + R6 = -1	R5 + B6 = 1
5 + 6 = 11	-5 + -6 = -11	5 + -6 = -1	-5 + 6 = 1

Set pupils the task of adding two numbers.

Set pupils the task of finding the difference between a negative number and a positive number and recording it as a measurement of temperature in ºC. Extend by drawing the range between the negative number and the positive number, horizontally or vertically. This task ensures pupils relate to the position of 0 and the recognition that -4 is greater than -6 etc.

It is possible to play **RRF!** using Black as positive and Red as negative, turning over two cards at a time and shouting the greater value, eg:

B2 R4   'two'
R3 R5   'minus 3'
B4 B8   'eight'

— —

Adding/subtracting near multiples of 10 to 2-digit numbers

Use a shuffled, digits deck to generate a 2-digit number. If the last card is red add 9; if black subtract 9. Demonstrate the technique adjusting multiples of 10:

'add 10, take away 1'   for add 9
'take away 10, add 1'   for take away 9

Extend more able pupils by increasing the 'near ten' multiples to 19, 29, etc by suit (eg if the last card is a heart + 9; a diamond + 19; a club -9; a spade -19).

Upper Key Stage 2

Operations showdown

For two or three players. Use a shuffled, full deck; face cards and Jokers have values 11, 12, 13 and 20. Deal five cards to each player. Players may use each operation (function) **only once**, to try to produce the highest total. The player with the highest score keeps the score. Repeat for a set number of rounds or set a **goal score**. The first player to reach or exceed it wins.

For example: 2, 9, 8, 4, 9   (9+8)x9-(4÷2)=151   but   (9+8)x9÷(4-2)=76.5

This can also be played with the lowest score winning – **golf** without the ball, just clubs!

Consolidate all times tables

Use a shuffled full deck and continue **RRF!**

Converting fractions to decimals to percentages

Using a shuffled, faceless deck, generate proper fractions and use a calculator to convert to decimals. Multiply the decimal by 100 to give a percentage. Multiplying by 100 should be done without the use of the calculator so that pupils practice moving numbers to the left of the decimal. Use the face cards, face down as zeros (place value holders).

$\frac{4}{5}$ is equal to 8 tenths or 80%

Deck Ahoy!

Adding proper fractions

Use a shuffled, faceless deck to generate two proper fractions. Look for a common denominator or create a common denominator. For example:

$$\frac{3}{4} + \frac{7}{8} \qquad \frac{6}{8} + \frac{7}{8}$$

$$\frac{2}{3} + \frac{5}{6} \qquad \frac{4}{6} + \frac{5}{6}$$

$$\frac{1}{3} + \frac{1}{4} \qquad \frac{4}{12} + \frac{3}{12}$$

It may be necessary to simplify the task by using the cards to generate only denominators and use '1' as the numerator – until pupils have grasped the concept of equivalent fractions and adding without confusion.

Demonstrate several examples and set the task of adding fractions generated by the cards. More able pupils could be set the task of adding three proper fractions.

Record and highlight all numerators greater than their denominators, to introduce the next activity.

- -

Improper (vulgar) fractions and mixed numbers

Use examples highlighted in the previous activity, eg $\frac{1}{4} + \frac{3}{4} + \frac{3}{4} = \frac{7}{4} = 1\frac{3}{4}$

to explain that when the numerator is greater than the denominator, the fraction is improper or vulgar. It can be converted into a mixed number by subtracting (or partitioning in diagrams) the equivalent of 1, eg 4 quarters.

- -

Subtracting fractions

Use a shuffled, faceless deck to generate two, proper fractions. Find or create their common denominator and equivalents. Write the fraction with the largest numerator first and subtract the other. Record and return the cards to the bottom of the deck. Repeat.

Doubling numbers with 1 decimal place

Use a shuffled, digit deck. Turn over three cards in a horizontal row. Turn the middle card face down to represent the decimal point. Record the number. Double the units. Double the tenths. Ensure pupils do not record 12 tenths as 12 hundredths. For example:

double 3.6 = 'double 3 is 6, double 6 tenths is 12 tenths or 1 and 2 tenths so double 3.6 is 7 and 2 tenths, 7.2'.

It may help some children to write the original number twice, in column form and add, carrying as in addition of whole numbers.

Extend more able pupils by using four cards eg 23.6, 45.9 etc. (Remember to keep one card face down as the decimal point.)

Halving numbers with 1 decimal place

As above. Ensure pupils understand that odd whole numbers will add 10 to their decimal value (ie add ten tenths.) For example: 'half of 2.8 is 1.4' but 'half of 3.8 is half of 2 plus half of 1.8', which is 1.9.

Divisibility tests

Although these have been covered before, they are worth revisiting with ever increasing numbers. Quick recognition of divisibility is a key skill in long division.

Test 2-digit numbers for divisibility. Test 3-digit numbers for divisibility. Use a shuffled deck, with the 10 and Jokers removed. The face cards can be turned down to represent zero.

Divisibility by 8

If the number is *even*, find half. If the answer is *even*, find half again. If the answer is *even*, the number is divisible by eight.

Divisibility by 9

Add the digits together. If the answer is a product of the 9x table, the number is divisible by 9. (*Or*, if it can be divided by 3 and the quotient can be divided by 3 again.)

Deck Ahoy!
© Janis A. Abbott and Brilliant Publications

Divisibility by 7

2–digit numbers greater than 70 can be 'tested' by subtracting the tenth multiple (ie 70.) If the remainder is a product in the 7x table, the number is divisible by 7.

Larger, 3-digit numbers can be 'tested' in the same way by subtracting multiples of the tenth multiple (**chunks** of 70), until the remainder, is identified as a product in the 7x table.

Larger 4-digit numbers can be 'tested' by partitioning 100s and 1000s (as tens of hundreds) and adjusted, increasing the tens/units component . For example:

4340   4300 + 40 = 4200 + 140

Both components are divisible by 7 as they are both multiples of the products 42 and 14.

Factors factory

Use a shuffled deck with the 10s and Jokers removed. The face cards can be turned down to represent 0. Generate 2-digit numbers (if 0 appears first, the number is single digit) and list all its factors using division facts or divisibility tests. Extend to 3-digit numbers or 4-digit numbers.

Extend by writing a list of possible fraction sentences. For example, 230 is divisible by 10, 5, and 2 so these fraction sentences are possible:

$\frac{1}{10}$ of 230 = $\frac{1}{5}$ of 230 = $\frac{1}{2}$ of 230 =

Dividing larger numbers by chunking

Use a shuffled, digit deck to generate 3-digit or 4-digit numbers. Record the number. Generate a 2-digit number as the divisor. Calculate x10 **chunks**, x5 **chunks** and x2 **chunks** before calculating. (Use a calculator to self-check the answer by multiplying the quotient by the divisor and adding any remainder. You should finish with the first 3-digit or 4-digit number generated.)

Multiples x10 showdown

Use a shuffled, full deck. (Face cards and Jokers have values 11, 12, 13 and 20.) Each player turns over two cards and multiplies each by 10. Then, each player multiplies their two multiples of 10. The player with the highest score wins all the cards. Vary by playing a set number of rounds or by setting a **goal score**.

Multiplying by 100/50/25

Use a stacked deck and single digits, initially, to introduce the concepts. Begin with x100, for the first suit. Extend to x50 for the second suit using the technique 'x 100 ÷ 2' (half) Repeat until secure. Extend to x25 using the technique, 'x 100 ÷ 4' (half and half again). Shuffle the deck when the pupil's knowledge is secure. **RRF!**

Extend to 2-digit numbers x100, x50 and x25.

— —

Multiply by multiples of 10

Use a stacked deck and single digits, initially, to introduce the concepts. Begin with x20, for the first deck, using the technique, x 2 (double it) x 10. Extend to x30, using the technique, x 3 (triple it) x 10 Extend to any multiple of 10.

Extend for more able pupils by using 2-digit and 3 digit numbers.

Shuffle the deck when the pupil's knowledge is secure. **RRF!**

— —

Adding numbers with decimals

This activity reinforces aligning decimals – so important! Use a shuffled, digit deck. Turn over three cards in a horizontal row. Repeat, placing the second row below the first row. Turn the middle card in each row, face down. This represents the decimal. Adjust the cards so the decimals are aligned and the numbers are in place value columns (units and tenths.) Record and add.

Extend for more able pupils. Turn over five cards in a horizontal row. Turn over another five cards in a horizontal row beneath. There should be two horizontal rows of five cards, face up. Turn down the 2nd or 3rd or 4th card in each row.

The face down card represents the decimal. Adjust to align the decimals *and* the place value columns. Record using zeros as *place value holders*. Add.

These decimal number activities can all be used with units of measurement to reinforce or extend calculations of length, area, weight, capacity and time.

Subtracting numbers with decimals

As *Adding numbers with decimals* (page 42), but record the number with the greatest value, above the lesser number. Then subtract.

– –

Multiplying with decimals

Multiply as though the numbers did not have decimals using any technique of partitioning and doubling, trebling, etc. Then simply apply the combined number of decimal places from the two original numbers, to the answer. For example:

2.6 x 7.1   (Each number is written to one decimal place so the answer will be written to two decimal places.)

26 x 71 = 1846
2.6 x 7.1 = 18.46   (*It works because* tenths x tenths equals hundredths, so the answer is written to 2 decimal places.)

2.63 x 7.1 = 18.673
263 x 71 = 18673   (One number is written to two decimal places, the other is written to one decimal place so the answer will be written to three decimal places. *It works because* hundredths time tenths equals thousandths.)

– –

Dividing with decimals

Dividing decimals is made easier by multiplying by 10s or 100s until the numbers are whole. Each number must be multiplied by the same multiple. So:

7.2 ÷ 0.3
72 ÷ 3 = 24   (Both numbers multiplied by 10 to give whole numbers.)

8.4 ÷ 0.12
840 ÷ 12 = 70   (Both numbers multiplied by 100 to give whole numbers.)

Task pupils to generate numbers to one or two decimal places using colour to differentiate the values. Using a digit deck, turn over 3 cards. If the last card is red, write the number to one decimal place. If the last card is black, write the number to two decimal places. Repeat. Finally, differentiate the operation: If the last card was even – multiply the two numbers. If the final card was odd, divide the larger number by the smaller number.

Calculating with mixed numbers

Use a digit deck to generate improper fractions, ie turn over two cards and use the highest value as the numerator. Convert the improper fraction into a mixed number. Repeat with two other cards so that you finish with two mixed numbers.

Model adding or subtracting mixed numbers by adding the integers, then the fractions. Demonstrate how this method is not always accurate for subtracting mixed numbers. Model the conversion of mixed numbers to improper fractions before subtracting. Reconvert to give an answer in mixed numbers.

Similarly, model the multiplication or division of mixed numbers by converting them to improper fractions, first.

- -

Multiplying fractions

Use a shuffled faceless deck or a full deck (for more able pupils). Turn over four cards and generate two proper fractions by using the lowest values as the numerators and the highest values as the denominators. For example:

5, 7, 2, 5 is written as $\frac{2}{5} \times \frac{5}{7}$

Write the number sentence. Multiply the two numerators. Multiply the two denominators. Write the final fraction. For example:
$\frac{2}{5} \times \frac{5}{7} = \frac{10}{35}$

More able pupils can be encouraged to look for a factor common to the final numerator and the final denominator, to reduce the fraction to its simplest form. For example:

$\frac{10}{35}$ is reduced to $\frac{2}{7}$ as both the numerator and denominator have been divided by 5.

The reduction by cancelling before calculating, is a skill for more able pupils. Remember to **cancel diagonally**.

$\frac{2}{5} \times \frac{5}{7}$ becomes $\frac{2}{1} \times \frac{1}{7}$ because the 5s have been cancelled diagonally.

Then **multiply horizontally** (ie numerator x numerator: denominator x denominator) to give the final answer: $\frac{2}{7}$.

Cancelling can take place with all parts of the fractions if reduction is possible. This is really for more able pupils at the Upper Key Stage 2 stage.

Deck Ahoy!

Dividing fractions

As above, generate two proper fractions. Write the number sentence by *inverting* the second fraction. Multiply as before! For example:

$$\frac{3}{4} \div \frac{1}{2} \quad = \quad \frac{3}{4} \times \frac{2}{1} = \frac{6}{4} = 1\frac{1}{2}$$

Continue to generate new fractions with four cards. (Pupils may need to review improper fractions and mixed numbers.)

— —

Operations showdown deluxe

This activity is for two or three players. Use a shuffled, digit deck. Deal five cards to each player. The black cards are positive numbers; the red cards are negative numbers. Each player tries to calculate the highest total of their cards, using each operation (function) *only once*! The player with the highest score wins all the cards or records all the scores or is the only player to score.

The rules can vary to suit the abilities and desires of the players. Extend, by setting a target of the lowest score *or* a score nearest to a target number. There are endless possibilities.

— —

Add the deck super-deluxe

Use a shuffled, digit deck and turn over three cards to make a 3-digit number. Turn over the next card. If it is black, multiply the 3-digit number by that score. If it is red, turn over yet another card to make a 2-digit number and add it to the original number. Keep a running score as you continue turning cards.

This can be played individually, collaboratively with a partner or competitively with a partner – the first player to correctly calculate wins a point. Players keep their own tally until the deck is complete!

Indices

Year 5 pupils are expected to recognize and calculate with squared and cubed numbers. The following activities allow them to stretch their minds a little further, too.

– –

Powers play

Use a shuffled, digit deck. Turn over a card and square it. Turn over the next card and square it. Find the sum of the two.

Variations

Include 'find the difference' or 'find the product'. Extend to cubed numbers. Mix the task according to colour, for example, red – square the number; black – cube the number.

– –

Powers play deluxe

Use a shuffled, digit deck. Turn over a card to represent the **base**. Turn over a second card to represent the **power** (exponent) and place it to the right and slightly above the first card. Calculate the **base** to the **power**. For example:

$5^7 = 5 \times 5 \times 5 \times 5 \times 5 \times 5 \times 5 = 78125$

– –

Probability

Use a faceless deck, initially. Start with only A, 2, 3 (or A, 2, 3, 4 ,5) if necessary.

Demonstrate the probability of selecting the A as 1/3 (or 1/5).

If the A is the first card selected, the probability of selecting A on the second draw is 0/2 (0/4) – impossible. Why? Because the remaining cards do not include an Ace and the total number of cards is now one less than before.

You can use the language of probability if preferred, ie probable, certain, impossible, etc. Pupils can be set the task of selecting any number of cards randomly from their own shuffled pack.

They can then record the probability of selecting each value, or suit, or picture as a fraction of their total number of cards.

Decide on a faceless or full deck according to the ability of the class.

© Janis A. Abbott and Brilliant Publications

Higher or lower

This is a popular and effective activity that teaches pupils to adjust the probability of an event in relation to previous events, using the language of probability.

Begin with one suit of cards. (It might be useful to have all the cards written down in numerical order so pupils can score them off as they appear and call *higher* or *lower* as they predict the probability of the next card to be turned.)

Establish the probability of each card being the first card turned face-up ie 1/13. Turn over the first card. Score it off the written list. Reconsider the probability of the next card being *higher* or *lower* in value to the first.

To play competitively in pairs or small groups, appoint one player as the dealer. Each player takes a turn to call *higher* or *lower* as a new card is turned. A correct call earns a point (perhaps a coloured counter). An incorrect call, doesn't earn a point. Once all the cards have been used, the player with the most counters has won!

— —

Ratio

The number of cards, selected randomly from a pre-set total, which are diamonds, for instance, could be used to pose differentiated problems for ratio. (It is unlikely that a random selection of cards will produce ratios of child-friendly multiples, so better to impose a simple ratio for children to use with their cards.)

'If the ratio of clubs to diamonds is 2:1 and you have _____ diamonds, how many clubs would you have?'

The actual number of diamonds in the pupil's selection, would be used to give a theoretical number of clubs.

'If the ratio of hearts to spades is 1:5 and you have _____ hearts' etc.

The actual number of hearts in the pupil's selection, would be used to give a theoretical number of spades.

Horatio

To play this competitive game, ratios are set before a game begins, awarding points according to the suit or colour of the turned card. For example: Red = 2:1 Black = 3:1.

If the first player scores the ratio of the first card, and the first card is a Red 6, they score 12 points. The second player scores on the second card. If the second card is a Black 5, they score 15 points.

Continue play until one player crosses the winning line (a score of 100, perhaps). The degree of difficulty of the ratios can be set according to the ability of the pupils. For example: an even card, ratio of 5:2 (so, divide the value by 2 and multiply by 5); an odd card, ratio of 4:3 (so divide by 3, multiply by 4) In this instance, evens would always score, but only 3, 6, 9 and Q would score! Quite challenging!

Deck Ahoy!

Statistics

These activities are suitable for Key Stage 1 and Lower Key Stage 2.

Sets

Model the drawing of two sets. Suggested labels could include:

Red/Black   Odds/Evens   High (>6)/Low (<7)

Use a full deck or a digit deck depending on the ability of your class or groups. Turn the cards and place in the correct set.

Extend to 2-digit numbers and change the parameters accordingly.

Extend further by introducing more demanding specifications, for example:

multiples of 3/even   multiples of 5/odd

This should introduce the possibility of some numbers meeting both specifications, ie they can be in both sets. Some numbers meet neither specification, ie they can not be in either set.

— —

Venn diagrams

Model the drawing of two sets overlapping to form an intersection. Ensure pupils understand that there are in fact four sets:
- ❖ set a
- ❖ set b
- ❖ intersection
- ❖ universal

Demonstrate using specifications that are familiar to all and place at least 1 number in each set. For example: multiples of 3/odd numbers

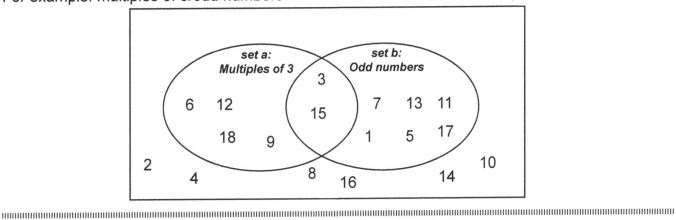

Graphs

Deck Ahoy! is great for representing sets of data in graphs, be they pictographs, column graphs or bar charts.

The colours and suits allow for different sets (and therefore columns and bars) at the simplest level. Pictographs are simple with ***Deck Ahoy!***: the diamonds and hearts in particular. The spade is usually drawn as an upside down heart with an isosceles triangle below; the club is usually drawn as three small circles with an isosceles triangle below.

Easier still, one card is represented by one rectangle. Alternatively, two cards could be represented by one rectangle and one card by half a rectangle, though it is easier to read if the division is diagonal, producing a right-angle triangle (kids don't draw half a rectangle accurately, in my experience!)

Turning a set number of cards (eg 30) means that differentiation is inherent because of the random selection.

Differentiation can be expanded by using different decks (eg more able pupils might use a full deck; others might use only a faceless deck) or by setting a range of numbers for each set or category (eg pupils might have to set their numbers according to $>_<$ parameters.) Sets from Venn Diagrams could also be used for graphing.

The variety of sets allows for a variety of column or bar graphs.

Extend to pie charts in Upper Key Stage 2.

Teaching time

Begin by discussing the terms analogue and digital. Stress the importance of the minutes and the relevant insignificance of the hours. The hour of the day can be estimated by light, shadows and the activity of living things eg birdsong, traffic, children playing etc.

The minute hand does the talking… so in our games of Clock solitaire we speak as the minute hand.

Clock solitaire

This is a traditional game, known worldwide I am sure, that I have adapted to help me teach children how to tell the time. Use all the cards except the Jokers. The Ace is 1, the Jack is 11, the Queen is 12 and the King is the centre point where the hands of the clock would pivot.

Shuffle the cards well and deal them (face down) onto a flat surface in the positions of the numbers on a clock – starting with 12, 6, 3, 9 and finally 1, 2, 4, 5, 7, 8, 10, 11, K.

Repeat until all the cards are used, (4 times in total) but turn over the last card as this will initiate the game.

Place the upturned card, face upwards, under the pile of cards at the position of that number. Turn over the top card of the pile and continue until all numbers appear face up. If you complete the clock before the Kings appear…*you win*! If the kings appear before all of the other numbers … *you lose*.

- —

Getting to know o'clock, half past, quarter past, quarter to

As you set out the clock in the recommended order – *only speak aloud* when placing the Queen ('o'clock'), the 6 ('half past'), the 3 ('quarter past') and the 9 ('quarter to'). Continue the game in the same manner, always repeating this language and *only* this language.

Extensions
Write the digital equivalent each time Q, 3, 6 and 9 appear.

Getting to know the 5-minute intervals on an analogue clock

Reinforce the 5x tables (up to 12 x 5 = 60). Direct pupils' attention to the minute intervals on the clock and highlight the 5-minute intervals at 1, 2, 3, 4, etc on the clock face. Set out the clock solitaire game exactly as before.

Before the game begins, reinforce the *past* and *to* sides of the clock. Use the medial line of the clock as a line of reflection. Demonstrate that *quarter past* is reflected as *quarter to*. Extend that language and concept to 5 *past* reflected as 5 *to*, 10 *past* reflected as 10 *to*, etc.

Play the game using all the language as above. Play **RRF!** in analogue time, For example:

8   twenty to
6   half past
Q   o'clock
5   twenty-five past

(Note: there is no need to specify am or pm or the *hour*, at this stage!)

— —

AM/PM

Play **RRF!** in analogue time where *red* cards are AM and *black* cards are PM. Use the current hour to complete the whole time sentence. For example:

Red 8   twenty to 10am
Black 3   a quarter past 10pm

Digital time and the 24-hour clock

Write the digital equivalents.

Chant the 5-minute intervals in digital time. Ensure that pupils understand that there are 60 minutes in an hour as calculated in the 5x table but in the language of time, we will never say '60'. Instead, we say 'o'clock' or '... Hundred hours'. Pupils seem to enjoy the reference to pilots' jargon:

'0 – 8 hundred hours'   for 8:00am
'13 hundred hours'   for 1:00pm

This gives an opportunity to introduce the 24 hour clock!

Play clock solitaire in the language of digital time.

Play **RRF!** using digital time. For example:

7   '35'
2   '10'
J   '55'
Q   'zero, zero'

Play **RRF!** where red cards are expressed in analogue time and black cards are expressed in digital time. (Pupils may wish to say any 24 hour clock expression they choose when they see a black queen. 'Zero hundred hours!' tends to be a favourite!)

— —

Telling the time to the minute

Play the game as before. Introduce four markers (plastic counters) three of one colour, one of another (eg 3 white, 1 red). These represent the minutes between the five minute intervals. Set the markers between two numerals on the *past* side of the clock face. The red marker is the minute the pupils must calculate. Express the analogue time. Repeat, changing the position of all four markers and the red marker!

Write the equivalent digital time on whiteboards.

Extensions

Play the game as before. Use the counters as before between two numerals on the *to* side of the clock face. Ensure pupils understand to count backwards from 30 to express the analogue time but count upwards from 30 to write the digital time!

Write the digital equivalents. Express the analogue time from the written digital time, ie as minutes *to the next hour*! (eg 15:42 is 18 minutes to 4).

Suggestions for further steps

❖ Introduce two hands to the clock (eg by attaching two straws or pipe cleaners of different lengths) and complete expressions with the hour indicated by the shorter hand.

❖ Set a start time and a finishing time and ask the pupils to calculate how much time has passed.

❖ Set a start time and state how much time is needed for a given activity. Ask pupils to calculate the finishing time (eg 'It will take 15 minutes to mix a cake and 25 minutes to bake it. It will need 20 minutes to cool. If it needs to be ready to eat at 6pm, what time will I need to start baking?')

❖ Set a finishing time and state how much time is needed for a given activity. Ask pupils to calculate the start time (eg 'It takes 25 minutes to walk to school. If the school bell rings at 08:55, what is the latest I can set off for school?')

❖ World time zones: play *Clock solitaire* with a partner and set each suit or colour to a different time zone!

Deck Ahoy!

Deck Ahoy! Club

It might seem incredible that pupils would attend a club after school to do yet more maths but *Deck Ahoy!* has attracted a dedicated following in my school. The beauty of the club is in watching the pupils teach each other or practice the activities they most enjoy.

'007' Ludo, *Take that snakes 'n' ladders* and *Clock solitaire* are always on the go! There is always a table set up with paper and pens for pupils to renew or refresh their *times tables mat.*

The *Deck Ahoy!* Club allows different ages to interact. The older children seem to enjoy teaching the younger pupils and the more able, younger pupils love to compete in *Showdowns* or *Darts* against the Y6s – especially their siblings!

Teachers and Teaching Assistants can see many different activities, and different levels, in action at the same time.

Parents sometimes attend (by open invitation) to clue-up on the homework!

It helps me, too! There is always an opportunity to conjure a new activity for a different year group!

And there you have it!

Not 'journey's end' because, I hope you will agree, there will always be another activity, another skill, to be mastered with the humble deck of cards.

Many thanks...

To the staff and pupils at:

All Saints' Junior School, Matlock, Derbyshire
Brassington Primary School, Derbyshire
Carsington And Hopton Primary School, Derbyshire
Duffield Meadows Primary School, Derbyshire
Kilburn Park School, London
Kirk Ireton Primary School, Derbyshire
Lyon Park Junior School, Wembley, Middlesex
South Wingfield Primary School, Derbyshire

Also to the many pupils I tutored privately, throughout the years and their parents.

My special thanks to the Head Teachers who at least tolerated and often encouraged my unorthodox methods, which have culminated in *Deck Ahoy!* They are: Mr Paul Addison, Mr Rex Bleakman, Mrs Rachel Bolton, Mr Peter Johnston, Mr Laurence Keel, Mrs Caroline Newton and Mrs Nicky Yudin.

I shall be forever grateful for your enthusiasm and support!